Nikon Z50 Camera

The Complete User Guide for Beginners and Seniors: Master Camera Settings, Lenses, Shooting Modes & 4K Video with Step-by-Step Tutorials, Real-Life Photography Tips, and Easy Troubleshooting

Georgette Howard

1

Table of Contents

Introduction

Just got your Nikon Z50? Feeling overwhelmed by buttons, settings, or blurry photos? This is the only beginner-friendly guide you'll need.

Whether you're a total beginner, a senior learning photography for the first time, or a smartphone user upgrading to a real camera, this step-by-step Nikon Z50 manual is your simple, supportive companion. Designed specifically for **first-time Nikon Z50 users**, this guide walks you through every essential feature from setup to shooting with clear language, helpful illustrations, and **real-world examples** that make learning fast, fun, and frustration-free.

What You'll Learn Inside:

- **Nikon Z50 for beginners and seniors**: Get started with confidence, even if you've never used a mirrorless camera before

- **Camera setup made easy**: Step-by-step instructions on charging, lens attachment, memory cards, buttons, menus & more

- **Every camera mode explained**: Auto, Program (P), Shutter (S), Aperture (A), Manual (M), and Scene modes

- **Mastering camera settings**: ISO, shutter speed, aperture, focus modes, and white balance simplified

- **Real-life photo recipes**: Capture sharp portraits, landscapes, motion shots, and indoor scenes without guesswork

- **4K video basics**: Learn how to record crisp, steady video for YouTube, family events, or travel vlogs

- **SnapBridge & Wi-Fi**: Transfer photos wirelessly and control your camera from your phone

- **Troubleshooting common issues**: Blurry images? Autofocus problems? Card errors? Solved.

- **Special tips for seniors**: Easy-to-read screen settings, larger buttons, and helpful grip adjustments

- **Creative challenges**: Fun photo ideas to try today—nature, pets, grandkids, holidays, and more

Ready to finally understand your camera and enjoy photography again?

This is the complete Nikon Z50 guide you've been searching for.

Part I: Getting Started with your Nikon

Chapter 1

Introduction to the Nikon Z50

Getting to Know the Camera That Will Change the Way You See the World

Why This Book is Different from the Nikon Manual

Let's be honest: the Nikon manual feels like it was written by engineers for robots. It's filled with jargon, diagrams that lack context, and instructions that assume you already know what aperture or ISO means. It's useful—yes—but only if you're already halfway to being a pro.

This book is **not** that.

This is your **real-world** companion. A friendly, easy-to-follow guide written for people who don't speak "camera tech"—people like you. Whether you're a senior picking up photography for the first time, a traveler looking to document your adventures, or someone upgrading from smartphone snapshots, this guide is here to walk you through everything, step by step, with plain English and real-life examples.

We'll not only explain what buttons to press, but also why, when, and how to use them in everyday situations. You'll feel confident using your Z50 to take better photos of your grandkids, vacations, garden, pets or anything else that matters to you.

Key Features of the Nikon Z50 (In Plain Language)

The Nikon Z50 is part of Nikon's new mirrorless camera lineup, which means it ditches the bulky mirror found in traditional DSLRs and replaces it with smarter, more compact technology.

Here's what makes it special:

- **Compact and Lightweight** – It's easy on the hands and won't weigh you down, which is great for travel and all-day use.

- **24.2 Megapixels** – That means sharp, detailed photos that still look great even when printed large.

- **4K Video Recording** – Capture crystal-clear videos with sound, perfect for birthdays, holidays, and personal projects.

- **Touchscreen Controls** – Like your smartphone, you can tap to focus, swipe to view photos, and pinch to zoom.

- **Eye-Detection Autofocus** – It automatically finds and focuses on people's eyes—ideal for portraits and family shots.

- **Built-in Wi-Fi and Bluetooth** – You can wirelessly transfer your pictures to your phone in seconds.

In short, the Z50 gives you professional-quality images in a body that's easy enough for a beginner to learn and grow with.

Mirrorless vs. DSLR: What It Means for You

You've probably heard the terms mirrorless and DSLR thrown around. Here's a simple way to understand the difference:

- **DSLRs** use a mirror system to reflect the image into an optical viewfinder.
- **Mirrorless cameras** like the Z50 don't have a mirror. Instead, they use electronic sensors and digital viewfinders.

What this means for you:

- **Lighter and Smaller:** Easier to carry.

- **Quieter:** No mirror "slap" when you take a photo.

- **More Accurate Previews:** What you see on the screen is exactly what the final image will look like.

- **Faster and Smarter Autofocus:** Great for capturing action or candid moments.

The Z50 keeps the quality of a DSLR but simplifies the shooting experience with modern, intuitive technology. It's like upgrading from a flip phone to a smartphone with all the power, but none of the bulk.

What's in the Box? Understanding Every Accessory

When you opened your Nikon Z50 box, you probably saw a mix of parts, cords, and paperwork and possibly had no idea what they all did. Don't worry, we'll decode it here:

Here's what typically comes in the standard Nikon Z50 kit:

- **Nikon Z50 Camera Body** – The heart of your system.

- **Nikkor Z DX 16-50mm Lens (or optional 50-250mm)** – Your general-purpose zoom lens.

- **Battery (EN-EL25)** – Rechargeable lithium-ion battery.

- **Battery Charger (MH-32)** – Plugs into the wall to charge your battery.

- **Camera Strap** – Clips onto the body so you don't drop your camera.

- **Body Cap** – Keeps dust out when no lens is attached.

- **Lens Caps (Front and Rear)** – Protect your lens from scratches.

- **USB Cable** – For transferring photos to a computer.

- **User Manual (Paper or Online)** – A technical reference (not fun to read).

Optional Accessories You May Also Have:

- **Memory Card (SD Card)** – Stores your photos. You can't shoot without it!

- **Camera Bag** – For protection and easy carrying.

- **Extra Battery** – Always helpful for longer shoots.

Throughout this guide, we'll show you how to use every part confidently. You won't need to guess.

Common Fears First-Time Users Have—And How We'll Solve Them

Most new camera users—and especially seniors—face a few common hurdles. If any of these sound familiar, you're not alone:

- "It's too complicated. I'll mess something up."

➤ You won't. This book breaks it down into small, simple steps with pictures and examples.

- "There are too many buttons. I don't know what they do."

➤ We'll cover every important button and ignore what you don't need right now.

- "I've never used a real camera before."

➤ Perfect. You're a blank slate and that's better than unlearning bad habits.

- "I just want to take good photos of my family, garden, or travels."

➤ That's exactly what we'll help you do quickly and confidently.

- "I don't want to read a textbook."

➤ Good news: This guide is conversational, visual, and friendly. You can skip around. No fluff.

By the end of this book, you won't just own a Nikon Z50, you'll understand it, enjoy it, and use it like a pro.

Ready? Let's get started.

Chapter 2

Your First Unboxing & Setup

Step-by-Step Instructions to Get Your Nikon Z50 Ready Without the Overwhelm

Attaching the Kit Lens the Right Way

Before you can take your first photo, you'll need to attach your lens to the camera body. If this is your first time doing it, take a deep breath, it's much easier than it looks.

Step-by-Step:

1. Remove the Body Cap

- Look at the front of the camera. Twist off the black plastic cap labeled "Nikon" (this protects the camera sensor).

2. Remove the Rear Lens Cap

- Turn your lens around and twist off the plastic cap covering the back part (the metal mount).

3. Align the Dots

- On both the lens and the camera, you'll see a white dot. Simply match the dots together and gently press the lens into the mount.

4. Twist to Lock

- Turn the lens clockwise (about a quarter turn) until it clicks. Now it's locked in.

TIP: Never force the lens. If it's not turning smoothly, double-check that the white dots are aligned.

Charging the Battery & Inserting Memory Cards

Battery Setup:

1. **Find the EN-EL25** battery in your box and the MH-32 charger.

2. Plug the charger into a wall socket and insert the battery (metal contacts first).

3. Let it charge fully (the light will turn off when done—about 2.5 hours).

4. Once charged, open the **bottom flap of your camera**, insert the battery with the gold contacts facing in, and push until it clicks.

Memory Card Setup:

1. Buy a **UHS-I SD** card, 64GB or higher recommended.

2. Open the same bottom compartment.

3. Slide the memory card in gently until it clicks (label facing the back of the camera).

4. Close the compartment.

TIP: Your camera won't shoot without a memory card, don't forget this step!

Navigating the Nikon Z50 Body: Every Button Explained (Visually)

Your Z50 might look intimidating with all its buttons and dials. But each one has a purpose and most of them, you won't even need to touch right away.

Here's a visual button map (insert diagram here with numbered labels). For now, focus on just a few:

Must-Know Buttons:

- **Shutter Button (Top Right)** – Press halfway to focus, all the way to take a photo.

- **On/Off Switch (Surrounding Shutter Button)** – Turn clockwise to power up.

- **Mode Dial (Top Left)** – Lets you switch between Auto, Manual, Scene modes, and more.

- **i Button (Back)** – Opens quick-access settings.

- **Playback Button (Arrow Icon)** – Lets you view your photos.

- **Menu Button (Top Left on Back)** – Opens your camera's full settings menu.

- **Touchscreen LCD** – Tap to focus, swipe through photos, pinch to zoom, or use it like your phone.

You'll grow familiar with more buttons over time, but this is all you need to get started.

How to Hold Your Camera Properly to Avoid Blurry Shots

Holding your camera the right way can make a huge

difference in photo sharpness especially if your hands shake slightly.

How to Hold Your Camera

Step-by-Step:

- **Grip the right side of the camera** with your right hand. Let your index finger rest on the shutter button.

- **Support the lens** with your left hand from underneath, don't let it dangle.

- **Tuck in your elbows** to your body. This gives you more stability.

- **Stand with your feet shoulder-width** apart to stay balanced.

- When possible, **use the viewfinder instead of the screen**. It brings the camera closer to your body and reduces shake

TIP: Breathe gently and exhale slowly as you press the shutter. It helps reduce movement.

Setting the Language, Time, and Date

When you first turn on the Nikon Z50, it'll prompt you to choose your basic setup preferences. Here's how to do it right:

Step-by-Step:

1. **Power on your camera** using the On/Off switch.

2. On screen, you'll see a **Language Selection** prompt. Use the arrow keys or touchscreen to choose your language (e.g., English).

3. Select your **Time Zone** (e.g., Lagos, London, Toronto).

4. Set the **Date Format** (Year/Month/Day is standard).

5. Adjust the **Date and Time** using the dials or touch controls.

6. Confirm with OK.

TIP: Setting the right date helps keep your photo files organized, especially when uploading or printing.

Updating Your Camera's Firmware

(Made Simple)

Firmware updates are like brain upgrades for your camera. Nikon occasionally releases updates to fix bugs or improve performance.

Step-by-Step:

1. Visit the official Nikon support page on your computer: https://downloadcenter.nikonimglib.com

2. Search for "Z50" and look for the latest firmware update.

3. Download the firmware file and save it to an empty, freshly formatted SD card.

4. Insert the SD card into your camera.

5. On your Z50, go to:

   - Menu → Setup Menu (wrench icon) → Firmware Version

6. If the camera detects a newer version, you'll see an option to **update**.

7. Select "Yes" and let the camera complete the update.

8. Do **not** turn off the camera during the process.

NOTE: Firmware updates aren't always required. Only update if you notice bugs or if the new version adds a feature you need.

You're Ready!

You've just completed the most important first steps: assembling, powering on, and preparing your Nikon Z50 for its first use. From here, you'll learn how to take your first photos, understand your camera's modes, and unlock the true power of photography—without the overwhelm.

Chapter 3

Nikon Z50 Menu Made Easy

Taking the Confusion Out of Settings—So You Can Focus on the Fun

What the Menu is Really For—and Why It's Not as Scary as It Looks

The word "menu" often scares beginners. It brings to mind long, scrolling lists and tiny icons that feel more like a puzzle than a tool. But once you understand what the menu on your Nikon Z50 actually does, it becomes your best friend.

Think of it like the **control center** of your camera. It doesn't take photos, it sets the stage for how your camera behaves when you do. From telling your Z50 how to focus, to choosing how your pictures are saved, the menu helps you customize your **camera to match your comfort zone and style.**

You don't need to know everything in the menu right away, just a few key settings that will immediately make your camera more intuitive and your photos more consistent.

And don't worry: this **guide won't overwhelm you**. You'll only learn the settings that actually matter. The rest? You can safely ignore for now.

Quick Setup: The 10 Menu Settings to Change Immediately

Before you dive into shooting, let's get your camera working the way you need it to.

Here are the **10 essential** settings every beginner should adjust first. You'll find all of these in the main Menu (press the **MENU** button on the back of your Z50).

1. **Image Quality → JPEG Fine**

- Why? This gives you great-quality photos without the hassle of editing RAW files.

2. **Image Size → Large**

- Ensures your photos are saved at the highest resolution.

3. **Auto ISO Sensitivity Control → On**

- Allows your camera to automatically adjust brightness in tricky lighting.

4. **ISO Sensitivity → 100**

- A good starting point for outdoor daylight shooting.

5. **Release Mode → Single Frame**

- One shot per button press—perfect for everyday photos.

6. AF-Area Mode → Auto-area AF

- Lets the camera choose the subject automatically. Great for beginners.

7. Vibration Reduction → ON (if lens supports it)

- Helps prevent blurry images, especially in low light.

8. Set Date & Time → Double-Check

- Ensures accurate file organization.

9. Auto Off Timers → Long

- Gives you more time before the screen dims or the camera powers down.

10. Beep → On

- Provides audible confirmation when focusing is complete—a great cue for seniors.

Pro Tip: You don't need to memorize these. Simply flip back to this list any time you reset or upgrade your camera.

Customizing Your My Menu Tab for Fast Access

Instead of digging through five tabs every time you need a setting, Nikon lets you create a custom menu with only the options you care about.

This saves time, avoids frustration, and makes the Z50 feel like your own.

Here's how to set it up:

1. Press the **MENU** button.
2. Scroll down to the **"MY MENU"** tab (green star icon).
3. Select **Add items.**

4. Choose frequently used options like:

- ISO Sensitivity Settings

- Image Quality

- Format Memory Card

- Reset Focus Point

- Silent Photography

- White Balance

- Release Mode

Now, whenever you press the MENU button and scroll to My Menu, all your most-used features are waiting for you—no hunting, no scrolling.

You can also reorder or remove items at any time under "My Menu settings."

Playback, Shooting, and Movie

Settings Explained Simply

The Z50 menu is organized into different colored sections. Here's a breakdown of what they do—and the few settings you'll actually want to care about in each.

PLAYBACK MENU (Blue Triangle Icon)

This controls how you view your photos after taking them.

- **Image Review → ON**

Automatically displays the photo you just took.

- **Rotate Tall → ON**

Keeps vertical shots upright when reviewing.

- **Playback Display Options**

Add helpful overlays like histogram, highlight alerts, and shooting data.

PHOTO SHOOTING MENU (Camera Icon - Red)

This is where you control how your photos are captured.

46

- **White Balance → AUTO**

Your camera will adjust to indoor/outdoor light automatically.

- **Metering → Matrix**

Best all-around option for beginners.

- **Flash Control → TTL (if flash attached)**

Keeps your flash exposures balanced and automatic.

MOVIE SHOOTING MENU (Camera Icon with Film Strip)

This controls your video recording preferences.

- **Frame Size/Rate → 1080p at 30fps or 4K at 24fps**

Choose based on your purpose. 1080p is perfect for casual use.

- **Movie ISO → Auto**

Lets the camera handle light levels while you record.

- **Wind Noise Reduction → ON**

Helps make your audio clearer when filming outdoors.

SETUP MENU (Wrench Icon)

This is where you manage all general camera operations and maintenance.

- **Format Memory Card**

Clears your card safely. Do this before every shoot (after backing up!).

- **Monitor Brightness → +1 or +2**

Easier to see in sunlight—especially helpful for seniors.

- **Firmware Version → Check Occasionally**

This is where you'll update if Nikon releases any new features or fixes.

Summary Snapshot:

If you feel overwhelmed:

- **Use Auto Mode** to begin

- **Set up My Menu** for speed

- **Learn just 10 settings** at the start

- **Ignore the rest** until you're ready to explore

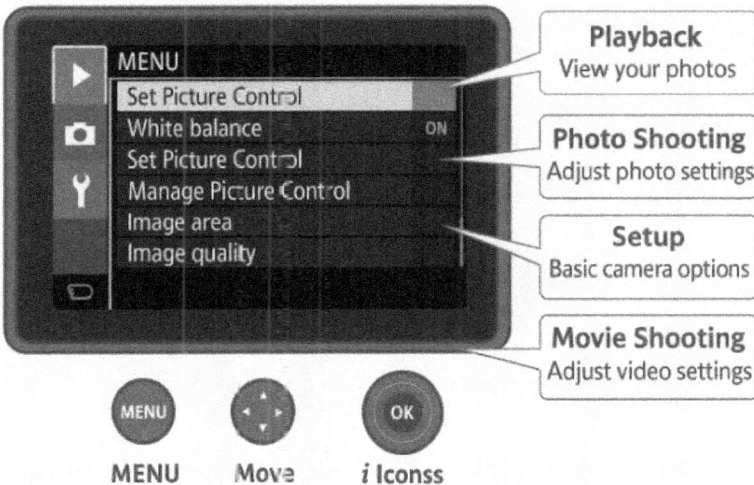

Nikon Z50 MENU MADE EASY

MENU

Set Picture Control
White balance   ON
Set Picture Control
Manage Picture Control
Image area
Image quality

Playback
View your photos

Photo Shooting
Adjust photo settings

Setup
Basic camera options

Movie Shooting
Adjust video settings

MENU   Move   *i* Iconss

Photography should feel exciting, not exhausting. The Nikon Z50 has tons of power but you don't need to master it all at once.

PART II: Camera Modes & Functions—Demystified

Chapter 4

Shooting Modes (Auto, P, S, A, M)

Mastering the Dial—So You Never Miss the Shot Again

Understanding Auto Mode—and When It Actually Fails

Let's start with the green Auto Mode on your mode dial. This is where most beginners start—and for good reason.

What Auto Mode Does:

- Chooses all settings for you: shutter speed, aperture, ISO, and white balance.

- Activates face detection and focus tracking.

- Turns on flash automatically when it thinks it's needed.

It's like cruise control: you point, shoot, and let the camera think for you.

But here's the catch:

- Auto Mode doesn't know what you're trying to *express*. It doesn't recognize:
- That you want a blurry background in a portrait.
- That you don't want the flash popping up during a candlelit dinner.
- That you're intentionally trying to freeze motion at a soccer game.

Bottom Line: Auto is perfect for everyday snapshots, but limiting when you're ready to get creative or shoot in tricky light.

Auto
Fully automatic shooting

P
Program mode for basic control

S
Shutter priority for motion

A
Aperture priority for backgrounds

M
Manual mode for full control

Program Mode (P): Best for Basic Control

Think of P Mode as "Auto with Options."

The camera still picks the shutter speed and aperture, but you can now adjust:

- ISO

- White balance

- Exposure compensation (make your image brighter or darker)

- Focus settings

- Flash on/off

Best Use Cases:

- Everyday shooting where you want a little creative say.

- Learning exposure slowly without being overwhelmed.

- Shooting in good lighting, both indoors and out.

Bonus: If you scroll the command dial, the camera rotates different combinations of aperture and shutter speed to keep exposure the same. It's like safe training wheels with room to experiment.

Shutter Priority (S): Capturing

Action Like a Pro

In Shutter Priority Mode, you set the shutter speed. The camera takes care of aperture.

Why Shutter Speed Matters:

- A **fast shutter** (like 1/1000 sec) freezes action.
- A **slow shutter** (like 1/30 sec) adds blur for motion or low-light.

When to Use It:

- Sports
- Wildlife
- Kids running around
- Waterfalls (slow shutter for silky effect)

Tip for Seniors: If your hands shake slightly, setting a faster shutter (like 1/125 or 1/250) can help reduce accidental blur.

Aperture Priority (A): Getting Those Blurry Backgrounds

In Aperture Priority Mode, you set the aperture, and the camera selects the best shutter speed.

Why Aperture Matters:

- A **wide aperture** (e.g. f/2.8) gives you soft, blurred backgrounds—perfect for portraits.
- A **narrow aperture** (e.g. f/11) keeps everything sharp—great for landscapes.

When to Use It:

- Portraits (use f/2.8–f/4 for beautiful depth)
- Landscapes (use f/8–f/11 for sharpness)
- Food or flower photography

This mode gives you creative **control over what's in focus and what fades away,** like painting with sharpness and blur.

Helpful Hint: Your Z50's kit lens ranges from f/3.5 to f/6.3 depending on zoom. Want more blur? Consider a prime lens like the Nikkor Z 35mm f/1.8.

Manual Mode (M): When You're Ready to Take Full Control

Manual Mode can feel intimidating but it's really just about owning the shot.

You decide:

- Shutter Speed
- Aperture
- ISO

When to Use It:

- Creative night photography (long exposures, light trails)

- Studio photography with controlled lighting

- Learning how all settings interact (great for photography learners)

- Difficult lighting where auto fails (e.g. stage lights, silhouettes)

Tip: Start by setting ISO to Auto even in Manual Mode. That way, you control exposure without stressing over brightness.

Manual Mode teaches you photography fundamentals faster than any course.

Scene Modes: Portrait, Night, Landscape, Sports & More

If full control feels too early, Nikon's **Scene Modes** offer a shortcut to great results.

You can find these by turning the mode dial to **SCENE** and selecting the type of photo you're taking:

Scene Options:

- **Portrait:** Smooth skin tones and blurred background
- **Night Portrait:** Balances subject with low-light background
- **Sports:** Fast shutter speed to freeze movement
- **Close-Up (Macro):** Great for flowers, food, or small details
- **Landscape:** Deep focus with enhanced greens and blues
- **Child:** Bright color and soft skin tones

Each mode auto-optimizes your settings without needing to touch menus or think technically.

Choosing the Right Mode for YOU

Situation	Recommended Mode
Everyday snapshots	Auto or P
Fast action (sports, kids)	S (Shutter Priority)
Portraits with blur	A (Aperture Priority)
Learning full control	M (Manual)
Simple creative presets	Scene

Quick Summary:

- **Auto** is great to start but knows nothing about your artistic goals.
- **P Mode** gives you flexibility without overwhelming you.
- **S Mode** = control motion and action.
- **A Mode** = control depth and focus.

- **M Mode** = control everything.

- **Scene Modes** = creative results with zero technical know-how.

Chapter 5

Understanding the Nikon Z50 Touchscreen & Viewfinder

Making the Most of What You See—and How You Interact with It

Using Touch to Focus, Shoot, and Swipe Through Images

Your Nikon Z50 touchscreen isn't just for navigating menus, it behaves much like a smartphone. You can tap, swipe, zoom, and even shoot with just a fingertip.

Here's what you can do:

- **Tap to Focus:**

In live view, simply tap on your subject on the LCD screen.

The camera will focus exactly where you tapped—perfect for portraits or off-center subjects.

- **Touch Shutter:**

Want to take a photo with a tap? Turn on "Touch Shutter" in the **i Menu**. Now, a tap anywhere on the screen will both focus and capture the shot. Great for tripod use or low-angle shots.

- **Swipe to View Photos:**

After shooting, press the Playback button and swipe left/right to scroll through images.

- **Pinch to Zoom:**

Use two fingers to zoom in on an image, just like on a phone.

- **Double Tap to Zoom In Quickly:**

Quickly see your subject's eyes or fine details without fumbling with buttons.

TIP: You can disable touch functionality in the settings if you find it too sensitive, but most users love its simplicity.

EVF vs LCD: What's Best for Your Eyes?

The Nikon Z50 offers **two ways to see your shot**:

1. **LCD Screen (back of the camera)**

2. **Electronic Viewfinder (EVF)**—the small eyepiece you look through

LCD Pros:

- Large and bright
- Easier to use for touch control and playback
- Better for composing wide shots or reviewing photos with others

EVF Pros:

- Blocks outside glare, great for shooting in bright sunlight
- Stabilizes the camera by pressing it to your face
- More comfortable for users with glasses (after adjusting the diopter)

Best Practice: Use the **EVF outdoors** and the **LCD indoors** or when tripod shooting.

Diopter Adjustment: Fixing Blurry Viewfinder for Seniors

Ever looked through the EVF and seen a fuzzy, unreadable image even though your LCD looks perfect? That's probably the **diopter** needing adjustment.

The diopter is a small dial next to the EVF that adjusts focus **for your vision**.

To Adjust:

1. Turn on your camera and look through the EVF.
2. Find the **diopter wheel** next to the viewfinder.
3. Look at the display text inside the EVF (not the subject).
4. Slowly rotate the dial left/right until the text appears sharp.

Wearing glasses? Adjust the diopter while wearing them. If you switch between glasses and no glasses, you may need to re-adjust.

This tiny dial can make a huge difference in comfort and clarity especially for older users.

Chapter 6

Buttons, Dials, and Real-World Usage

Getting Hands-On With Your Z50—One Control at a Time

What Each Button Does—and How to Remember It Easily

The Z50 isn't overloaded with buttons, but if you're new to digital cameras, they can still feel mysterious. Let's break down the essentials.

Top of the Camera:

- **Mode Dial** – Switch between Auto, P, S, A, M, Scene, and Video

- **Shutter Button** – Press halfway to focus, all the way to shoot

- **On/Off Switch** – Encircles the shutter button

- **Exposure Compensation (+/-)** – Brighten or darken your photo manually

- **Video Record Button (Red Dot)** – Starts and stops video

Back of the Camera:

- **Menu Button** – Opens the full camera menu

- **i Button** – Quick settings overlay (customizable)

- **Playback Button** – View your photos and videos

- **Delete Button (Trash Icon)** – Remove unwanted shots

- **Multi-selector (Arrow Pad)** – Navigate menus and move focus points

- **OK Button** – Confirm selections

- **Zoom In/Out Buttons** – Helpful in playback and live view

- **Display (DISP) Button** – Toggles what info appears on screen

Front of the Camera:

- **Lens Release Button** – Detach the lens

- **Function Button (Fn)** – Can be customized (e.g. for ISO or White Balance)

How to Use the Command and Sub-

Command Dials

The Z50 has **two dials** for controlling settings without diving into the menu

- **Main Command Dial (Rear Dial):**

Usually controls the **shutter speed** or browsing through images

- **Sub-command Dial (Front Dial near grip):**

Controls **aperture** or additional settings depending on the mode

Example:

- In **Manual Mode**, rear dial = shutter speed, front dial = aperture
- In **Aperture Priority**, only the front dial is active

TIP: Use these dials while looking at the screen to see immediate changes—this builds muscle memory fast.

Top 10 Most Used Buttons and Their Real Purpose

Button	Purpose	When to Use
Shutter	Take photo	Always
Mode Dial	Choose shooting mode	Before each shoot
i Button	Quick settings access	Anytime
Playback	View photos/videos	After shooting
Zoom In	Check photo sharpness	After shooting
Delete	Remove bad	After review

		shots
Menu	Access full settings	Setup or advanced options
DISP	Simplify screen clutter	While composing shots
AF-On (optional config)	Manual focus control	Action shots
Fn Button	Customizable	Set for ISO, WB, or silent shooting

Final Thought:

Once you learn your Z50's physical layout, it stops being "a camera" and starts becoming an extension of your hand and eye. With practice, you'll be switching modes, adjusting brightness, and capturing beautiful moments— all without needing to stop and think.

PART III: Photography Fundamentals Made Simple

Chapter 7

Lenses, Zooming & Focusing

Everything You Need to Know to Choose the Right Lens and Nail the Focus Every Time

Choosing the Right Lens (Kit Lens vs Prime vs Telephoto)

The Nikon Z50 uses **Nikon Z-mount lenses**, specifically the **DX-format Z lenses**. If you bought the camera in a kit, you likely received one or two lenses already.

Let's break down the main types:

Kit Lens (NIKKOR Z DX 16–50mm f/3.5–6.3 VR)

This is the lens most beginners start with and it's actually very versatile.

- **Zoom range:** Great for landscapes, street shots, group photos.

- **Compact and lightweight:** Perfect for travel.

- **VR (Vibration Reduction):** Helps reduce blur from hand shake.

- **Wide to normal field of view:** Good for everyday photography.

Ideal for: Learning, casual use, city walks, family outings.

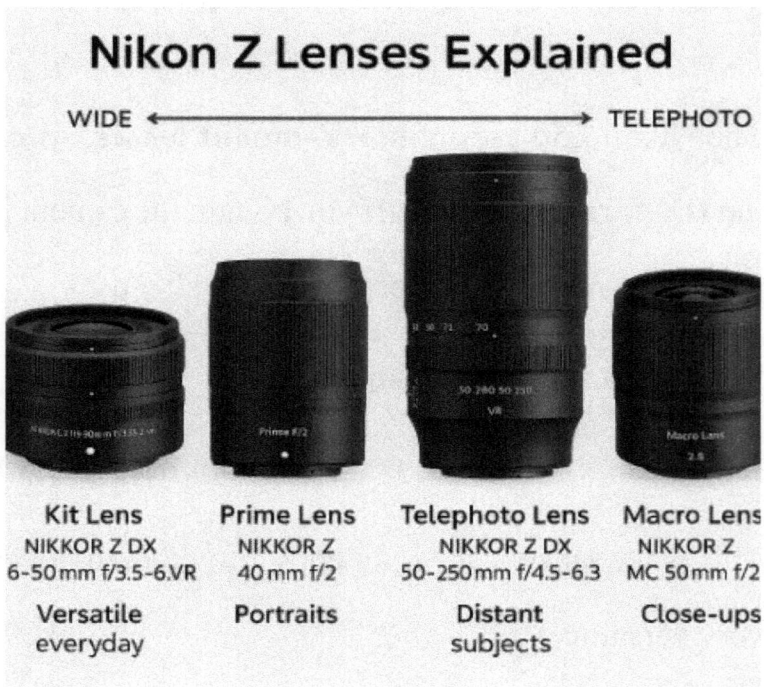

Nikon Z Lenses Explained

Kit Lens	Prime Lens	Telephoto Lens	Macro Lens
NIKKOR Z DX	NIKKOR Z	NIKKOR Z DX	NIKKOR Z
6-50mm f/3.5-6.VR	40mm f/2	50-250mm f/4.5-6.3	MC 50mm f/2
Versatile everyday	Portraits	Distant subjects	Close-ups

Prime Lens (e.g. NIKKOR Z 40mm f/2 or 35mm f/1.8)

A prime lens has no zoom. You move your feet to "zoom" in or out.

- **Sharper images** and **better low-light performance**
- **Creamy background blur (bokeh)** ideal for portraits
- **Lightweight and fast** focusing

Ideal for: Portraits, food photography, low light, creative control.

Telephoto Lens (e.g. NIKKOR Z DX 50–250mm f/4.5– 6.3 VR)

This lens brings **faraway subjects closer**.

- **Great for wildlife, sports, or concerts**
- Has **VR stabilization** to help reduce shake
- Bulky, but powerful

Ideal for: Outdoor shooting, travel, nature, action shots.

Manual Focus vs Autofocus: What's Better for You?

The Nikon Z50 has an excellent autofocus (AF) system. But there are times when you may want to switch to manual focus (MF).

Autofocus (AF)

- Fast, smart, and works great in most lighting
- Use **Face/Eye Detection** for portraits
- Works with touch screen (tap to focus)
- Perfect for beginners, travel, and general photography

Great for: Everyday shooting, quick snapshots, video, moving subjects

Manual Focus (MF)

- You control the focus by turning the ring on your lens

- Essential for **macro photography**, **low-light scenes**, or **fine adjustments**

- Use **Focus Peaking** to highlight what's in sharpest focus

Best for: Learning focus, slow compositions, creative control

TIP: Switch between AF and MF in the camera's i Menu or side lens switch (if available).

Using Nikon's Silent Autofocus for Videos or Quiet Settings

Ever tried recording a video only to hear click-click-click from your camera adjusting focus?

The **Z50's stepping motor (STM)** lenses, like the 16–

50mm kit lens—offer near-silent autofocus that's **perfect for video recording** and quiet environments (e.g. weddings, libraries).

Here's how to make the most of it:

Silent Autofocus Setup:

1. **Use a Z DX lens** with STM (like 16–50mm or 50–250mm).
2. Set **AF-F (Full-time AF)** when in video mode.
3. Enable **Face Detection AF** if recording people.
4. Set **Movie Sound Level to Auto** or adjust manually to avoid over-recording focus noise.

Ideal for vloggers, interviews, nature, and intimate moments.

Understanding Lens Markings (e.g.,

16–50mm, f/3.5–6.3)

Your lens has numbers and symbols that may look like code. Let's decode them:

Example: NIKKOR Z DX 16–50mm f/3.5–6.3 VR

Marking	Meaning
16–50mm	Zoom range—16mm = wide angle, 50mm = normal view
f/3.5–6.3	Maximum aperture at different zooms. f/3.5 (at 16mm) lets in more light than f/6.3 (at 50mm)
VR	Vibration Reduction—helps reduce blur from hand shake
DX	Designed for APS-C (cropped sensor) cameras like your Z50

Wider numbers (like 16mm) are great for landscapes and groups. Longer numbers (like 50mm or 250mm) zoom in on faraway subjects.

Summary:

- **Start with your kit lens:** It's more powerful than it looks.

- **Try a prime lens** if you want sharper photos or blurry backgrounds.

- **Use autofocus** for simplicity; switch to manual when you need precision.

- **Enable silent AF** for quiet, professional-looking video.

- **Learn to read lens numbers**—they're the key to understanding what your lens can really do.

Chapter 8

Understanding ISO, Shutter Speed, and Aperture

Master the Exposure Triangle—Without the Headache

Plain English Explanation of the Exposure Triangle

Every photo is created by **light** entering your camera. Your Nikon Z50 controls this light using **three main settings**, often called the **Exposure Triangle**:

1. **ISO** – Controls the camera's sensitivity to light.
2. **Shutter Speed** – Controls how long light enters the camera.
3. **Aperture (f-stop)** – Controls how wide the opening is for light to pass through.

Imagine your camera as a window:

- **ISO** = How dirty or clean the window is (clean = low ISO, dirty = high ISO).

- **Shutter Speed** = How long the window is open.

- **Aperture** = How wide the window opens.

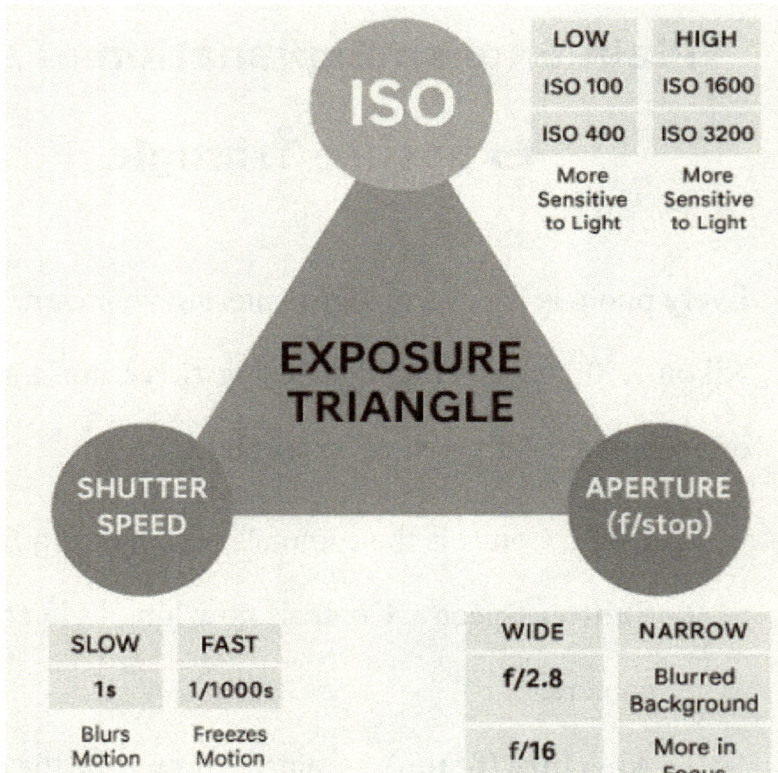

	LOW	HIGH
ISO	ISO 100	ISO 1600
	ISO 400	ISO 3200
	More Sensitive to Light	More Sensitive to Light

EXPOSURE TRIANGLE

SHUTTER SPEED

APERTURE (f/stop)

SLOW	FAST
1s	1/1000s
Blurs Motion	Freezes Motion

WIDE	NARROW
f/2.8	Blurred Background
f/16	More in Focus

Real-World Analogy: Filling a Glass with Light

Imagine you're filling a glass of water:

- **Aperture = Faucet Size**

A wide-open faucet (f/2.8) lets in water quickly. A narrow faucet (f/11) takes longer.

- **Shutter Speed = Time the Faucet is Open**

Open the faucet for 1 second vs 1/1000 of a second.

- **ISO = Water Filter Sensitivity**

A clean filter (ISO 100) gives pure water but works slowly. A rough filter (ISO 3200) fills faster but may add noise or dirt.

Goal: Fill the glass just right—not too much, not too little—every time you take a photo.

How These Settings Affect Your

Photos

Let's break it down visually and practically:

ISO: Brightness & Image Quality

ISO Setting	Light Sensitivity	Image Quality
ISO 100	Low sensitivity	Crisp, no grain
ISO 400	Moderate	Good for daylight
ISO 1600	High sensitivity	Slight grain, okay for indoor use
ISO 3200+	Very high	Grainy, use in dark scenes only

TIP: Keep ISO low (100–400) in good light for best image

quality.

Shutter Speed: Freezing or Blurring Motion

Shutter Speed	What It Does	Example Use
1/1000 sec	Freezes fast action	Sports, wildlife
1/250 sec	Everyday moments	Portraits, kids
1/60 sec	Standard handheld	General scenes
1/10 sec	Adds motion blur	Light trails, creative shots

TIP: Use a tripod or VR lens if going below 1/60 sec to avoid blur from shaky hands.

Aperture (f/stop): Focus Depth & Light

Aperture	Opening Size	Focus Effect	Use
f/2.8	Wide	Blurry background (shallow depth)	Portraits
f/5.6	Medium	Balanced blur and sharpness	Everyday
f/8–f/11	Narrow	Everything in focus (deep depth)	Landscapes

TIP: Lower f-number = more background blur. Higher f-number = sharper across the whole image.

Step-by-Step Examples: What

Settings to Use

Let's apply this triangle to real photography scenarios.

- **Portraits (Soft Background)**

- **Aperture:** f/2.8 or f/4

- **Shutter Speed:** 1/125 or faster

- **ISO:** 100–400

- **Tip:** Use Eye-Detection AF and shoot with natural light

Landscapes (Sharp Focus)

- **Aperture:** f/8 to f/11

- **Shutter Speed:** 1/125 (tripod optional)

- **ISO:** 100

- **Tip:** Use manual focus or focus on a point 1/3 into the scene

Action (Sports, Pets, Kids)

- **Aperture:** f/4 to f/5.6

- **Shutter Speed:** 1/1000 or faster

- **ISO:** 800–1600 (depending on light)

- **Tip:** Use Continuous AF and burst mode

Night Scenes

- **Aperture:** f/2.8 or wider

- **Shutter Speed:** 1/10 to 10 seconds (tripod needed)

- **ISO:** 1600–3200

- **Tip:** Turn off flash for more natural light effects

Summary:

- ISO = Brightness **sensitivity**

- Shutter Speed = Motion **control**

- Aperture = Focus and **depth**

Balance all three to get beautiful, intentional photos—even in tricky lighting.

Chapter 9

Real Photography Scenarios & Walkthroughs

Shoot with Purpose—And Get the Shot You Want, Every Time

Taking a Sharp Portrait with Beautiful Background Blur

Nothing feels more rewarding than capturing a portrait where the subject pops and the background melts into softness.

Best Settings for Portraits:

- **Mode:** Aperture Priority (A)
- **Aperture:** f/2.8 – f/4 (as low as your lens allows)
- **ISO:** 100–400
- **Shutter Speed:** Auto (camera will adjust)

- **Focus Mode:** Single-point AF (AF-S) or Eye-Detection AF
- **Lens:** Prime lens (like 35mm or 50mm) or kit lens at 50mm zoom

Step-by-Step:

1. Set your mode dial to **A**.
2. Rotate the front dial to your **widest aperture** (lowest f-number).
3. Enable **Eye-Detection AF** in the i Menu.
4. Ask your subject to step 5–10 feet from the background.
5. Focus on their eye—hold steady—and shoot.

*TIP: Shoot at **eye level**. Use **natural light** from a window or outdoors for a soft, flattering look.*

Capturing Pets or Children in

Motion

Wiggly kids or hyper pups won't wait for your camera. You need speed, tracking, and burst mode.

Best Settings:

- **Mode:** Shutter Priority (S)
- **Shutter Speed:** 1/1000 sec or faster
- **ISO:** Auto or 400–800
- **Focus Mode:** Continuous AF (AF-C)
- **AF-Area Mode:** Dynamic or Auto-Area AF
- **Release Mode:** Continuous (Burst)

Step-by-Step:

1. Turn the mode dial to **S**.
2. Use the rear dial to set **1/1000 sec**.
3. Tap **i Button → AF-C + Auto-Area AF**.
4. Enable **burst mode** (press left on multi-selector).

5. Focus, hold shutter halfway to track movement, then shoot in bursts.

💡 *TIP: Get low to their eye level, use natural light, and **let the moment unfold**—don't pose, just follow.*

Real Photography Scenarios & Walkthroughs

Taking a sharp portrait with background blur
Aperture Priority (A)
f/2.8
ISO 100

Capturing pets or children in rmotion
Shutter Priority (S)
1/1000 s
Auto

Shooting sunsets and low-light scenes
Aperture Priority (A)
f/8
ISO 100 (tripod)

Travel photography: city, food, architecture
Program Auto
Auto ISO

Shooting Sunsets and Low-Light

Scenes

Low light brings mood, color, and drama—but it requires careful settings to avoid blur and grain.

Best Settings:

- **Mode:** Aperture Priority (A) or Manual (M)
- **Aperture:** f/5.6 – f/8
- **ISO:** 100–400 (tripod) or 800–1600 (handheld)
- **Shutter Speed:** Let the camera decide (or 1/60–1 sec in Manual)
- **White Balance:** Cloudy (warmer tones)
- **Lens:** Wide lens (16–24mm range)

Step-by-Step:

1. Use **A mode** and set to f/**8.**
2. Mount your camera on a **tripod** (or brace against a wall).

3. Set ISO to **100** and use **2-second self-timer** to avoid shake.

4. Compose so the sun is off-center for balance.

5. Review and adjust exposure compensation if needed.

TIP: For deeper color, **underexpose slightly** *(lower exposure compensation to -0.3 or -0.7).*

Indoor Photography Without Flash

No one likes the harsh white burst of a flash indoors. Use soft lighting and tweak your settings for a natural look.

Best Settings:

- **Mode:** Aperture Priority (A)

- **Aperture:** f/2.8 – f/4

- **ISO:** 800–1600 (depending on light)

- **Shutter Speed:** Camera will adjust

- **White Balance:** Auto or adjust to light type (e.g., incandescent)

- **Vibration Reduction:** ON

Step-by-Step:

1. Choose **A mode** and widest available aperture.

2. Increase ISO to **800 or 1600**.

3. Use available window light or soft lamps (avoid overhead).

4. Steady your grip—brace your elbows.

5. Avoid direct backlight unless you want a silhouette.

TIP: Stand where light falls onto the subject's face, not behind them, and always watch for unwanted shadows.

Travel Photography: City, Food, Architecture

When traveling, you want variety, speed, and control

without constantly changing settings.

Best Settings:

- **Mode:** Program (P) or Aperture Priority (A)

- **Aperture:** f/5.6 for general shooting

- **ISO:** Auto (max ISO 1600–3200)

- **Shutter Speed:** Auto

- **Lens:** Kit lens or a compact zoom

Tips by Subject:

City Life:

- Use **A mode**, f/5.6, Auto ISO.

- Capture people, movement, markets—don't wait for perfection.

Food:

- Use A mode, f/2.8–f/4.

- Shoot from a 45° angle or directly above (flat lay).

- Focus on textures and details.

Architecture:

- Use f/8 for sharpness.

- Turn on grid lines to align your shot.

- Capture both wide shots and close-up textures.

TIP: Keep your camera accessible. Travel photography is about reaction and curiosity, not perfection.

Summary:

- Use **Aperture Priority for control**, **Shutter Priority** for action, and **Manual for creativity**.

- Choose the **right ISO** and **aperture** for the scene.

- Lean on your camera's tools: Eye AF, VR, Silent Mode, Auto ISO.

- Practice in real settings—grocery store, park, living room—to build confidence fast.

PART IV: Mastering Video & Vlogging on the Z50

Chapter 10

Video Settings for Beginners

Capture Life in Motion—Without Confusion or Complicated Gear

How to Switch to Movie Mode

Switching your Nikon Z50 into video mode is as simple as turning a dial and pressing a button. You don't need a special lens, and there's no app required.

Step-by-Step:

Turn the mode dial to the **Movie Camera Icon**.

- Look at the **LCD screen or viewfinder**—your live view is now set for video.

- Press the **red Movie Record button** (next to the shutter) to start and stop recording.

TIP: If you're in any photo mode (like A or Auto), you can still record short clips by pressing the red record button but switching to dedicated video mode gives you full control.

Best Video Settings for YouTube, Social Media, and Home Movies

The Nikon Z50 makes shooting great-looking videos easy, but the right settings can make your footage shine especially for platforms like YouTube, Instagram, and Facebook.

For YouTube or Talking-Head Videos:

- **Resolution:** 4K UHD
- **Frame Rate:** 30fps
- **AF Mode:** AF-F (Full-Time Focus)
- **Sound:** Auto-level or plug in a microphone

- **Stabilization:** Enable VR (in-lens vibration reduction)

*Use an **external mic** for better audio if you're filming yourself speaking.*

For Instagram Reels or TikTok:

- **Resolution:** 1080p Full HD
- **Frame Rate:** 60fps (for smooth slow-mo or handheld shooting)
- **Vertical Orientation:** Flip the camera vertically (video editing apps will crop if needed)
- **Autofocus:** Use Face Detection or Subject Tracking

TIP: 60fps is better for slow motion. You can slow it down during editing for dramatic or fun effects.

For Home Movies or Everyday Moments:

- **Resolution:** 1080p (saves space, still great quality)
- **Frame Rate:** 30fps
- **ISO:** Auto

- **Focus Mode:** Face or Eye Detection

- **Use VR:** Helps avoid shaky footage

VIDEO SETTINGS
FOR BEGINNERS

Switch to Movie Mode
- Turn the mode dial to camera icon, then press the red Record button to start or stop video recording

Best Video Settings
- 4K at 30fps
  For YouTube
- 1080p at 60fps
  For Instagram
- 1080p at 30fps
  For everyday video

Using Autofocus During Video
- Set Focus mode to AF-F for continuous autdocus
  Tap the screen to

4K vs 1080p
- Sharper image
- Larger file size

1080p

Using Autofocus During Video

The Nikon Z50 features silent, smooth autofocus designed for video. This makes a huge difference especially when your subject is moving.

Autofocus Setup:

1. In video mode, press the **i Button**.

2. Set **AF Mode to AF-F (Full-Time AF).**

3. Choose **AF Area Mode → Face Detection AF** for people, or **Wide-Area AF** for general scenes.

4. Use **Touch Focus** on the LCD to switch focus between subjects while filming.

Silent AF is especially helpful when using the 16–50mm STM kit lens, which has a quiet motor and minimizes lens noise.

Pro Tips:

- **Half-press the shutter** to refocus manually without stopping the video.

- Turn on **Focus Peaking** if switching to manual focus for artistic control.

- Use **a tripod or handheld grip** for stability when filming family events or travel.

Understanding 4K vs 1080p (And Which to Choose)

When it comes to video resolution, you've got two primary choices:

4K UHD (3840 × 2160)

- Ultra-sharp, professional-looking
- Ideal for YouTube or large-screen playback
- Larger file sizes
- Crops the frame slightly when recording

*Use **4K** when video quality matters most.*

1080p Full HD (1920 × 1080)

- Standard quality for most casual video
- Smaller file size, longer recording time
- No cropping or sensor limitations
- Ideal for social media or everyday videos

Use **1080p** for quick sharing, vlogs, and memory capturing.

Resolution Summary:

Setting	Best For	Pros	Cons
4K (30fps)	YouTube, pro projects	Crisp detail	Big file size
1080p (60fps)	Instagram, slow motion	Smooth motion	Slightly less sharp
1080p (30fps)	Family videos, everyday	Balanced	Standard quality

Final Thoughts:

You don't need to be a filmmaker to shoot great video with your Nikon Z50. Just remember:

- **Start in Auto Movie Mode**
- Use **Face/Eye AF** for people

- Stick with **1080p at 30fps** for most everyday filming

- **Practice** small clips first, then build confidence

Your Z50 is a storytelling tool. With just a little setup, you can turn everyday moments into memories worth watching again and again.

Chapter 11

Real-Life Video Use

Capture the Sounds and Stories That Matter—Easily and Beautifully

Recording Family Moments

Whether it's a birthday, a toddler's first steps, or grandma telling a story, the Z50 is perfect for filming precious memories in natural, lifelike detail.

Recommended Settings:

- **Mode Dial:** Movie mode (camera icon)
- **Resolution:** 1080p at 30fps (perfect for everyday video)
- **Focus Mode:** AF-F (Full-time autofocus)
- **AF-Area Mode:** Face Detection

- **Microphone:** Built-in (or optional external mic)

- **Stabilization:** VR ON (if lens supports it)

Step-by-Step Guide:

1. Switch to **movie mode** and turn on your Z50.

2. Use the **flip-down screen** to check composition (especially when filming children).

3. Tap the **face** on the screen to focus.

4. Press the **red movie record** button to begin filming.

5. Let the moment unfold naturally, don't interrupt with prompts unless needed.

TIP: Use a tripod or small tabletop stand to keep your hands free and stabilize long clips. Always shoot horizontal unless it's for Instagram stories or TikTok.

Interview Setup for Vlogging

If you're recording an interview, tutorial, or vlog (whether solo or with someone else), good sound and framing are key.

Ideal Settings:

- **Resolution:** 4K at 30fps (YouTube quality)
- **Audio:** External mic strongly recommended
- **Lens:** 16–50mm for wide framing
- **AF Mode:** AF-F with Eye Detection ON
- **White Balance:** Auto or Daylight (avoid yellow tones indoors)

Vlogging Setup Instructions:

1. Mount the camera on a **tripod** or **gorillapod**.
2. Plug in a **lapel** mic or shotgun mic (more on this below).
3. Sit about **3–6 feet from the camera.**
4. Set to **AF-F** with Eye Detection enabled.
5. Use **Manual Exposure** or **Aperture Priority (f/4– f/5.6)** for flattering focus.

6. Check framing and audio before hitting record.

*TIP: Add **soft lighting** from a window or ring light. Record a **5-second** test clip and play it back to check focus, brightness, and sound before diving into your message.*

REAL-LIFE VIDEO USE

Recording family moments

Interview setup for vlogging

Connecting external microphone for better sound

Connecting External Microphones for Better Sound

Great video needs great audio. The Nikon Z50's built-in mic works fine but external microphones dramatically improve sound clarity, especially for:

- Interviews

- Vlogs

- Indoor recordings

- Noisy outdoor environments

Microphone Types & Recommendations:

Mic Type	Best For	Example
Lavalier Mic (clip-on)	Interviews, speaking to camera	BOYA BY-M1, Rode Lavalier Go
Shotgun Mic	Vlogging, talking from behind camera	Rode VideoMicro, Deity V-Mic D3
Wireless Mic	Movement or interviews at	Rode Wireless GO II, DJI Mic

distance

How to Connect:

1. Plug the mic's **3.5mm audio cable** into the **MIC input** on the left side of the camera (flip open the rubber flap).

2. The camera will **automatically switch** from built-in to external audio.

3. Press **MENU** → **Movie Settings** → **Microphone sensitivity**, and test levels (keep audio in the green zone, not red).

4. Optional: Enable **Headphone monitoring** via external recorder for live audio feedback.

Pro Audio Tips:

- **Avoid recording in echoey rooms**. Add rugs, curtains, or pillows to absorb sound.

- Use **Auto sound level**, or manually adjust to avoid distortion.

- Watch out for **wind noise outdoors**—use a **dead cat windscreen** if needed.

Summary:

- **For family moments**, use Full HD, Face Detection, and let things flow naturally.

- **For vlogs and interviews**, stabilize your camera, use external mics, and keep eye detection ON.

- **For best sound**, invest in a simple lav or shotgun mic and monitor your audio levels before recording.

PART V: Custom Settings, Maintenance & Troubleshooting

Chapter 12

Customizing Your Camera for Comfort

Make Your Z50 Easier to Use—Your Way, Your Style

Assigning Buttons for Faster Access

The Nikon Z50 lets you assign common tasks like ISO adjustment or white balance to buttons you can reach easily. This saves time, especially when you're outdoors, filming, or shooting moving subjects.

How to Assign a Button:

1. Press the **MENU** button.

2. Go to **Custom Setting Menu** (pencil icon).

3. Scroll to **Controls → Custom control assignment**.

4. Choose a button to customize, like:

   - **Fn1 (front lower button)**

- **Fn2 (front upper button)**

- **AE-L / AF-L** button (back right)

5. Select the function you want to assign (e.g., ISO, White Balance, Subject Tracking, etc.).

6. Press OK to save.

Recommended button shortcuts for beginners:

- **Fn1** → ISO Sensitivity

- **Fn2** → White Balance

- **AE-L / AF-L** → Subject Tracking

TIP: Use muscle memory—assign frequently used settings to where your fingers naturally rest.

Creating Custom Shooting Modes

Let's say you love specific settings for:

- Portraits with blurry backgrounds

- Action shots of your dog

- Vlogging in your room

Instead of resetting everything each time, you can save these setups to a U1 or U2 preset.

How to Save a Custom Shooting Mode:

1. Set your camera with the desired settings (e.g., mode, aperture, ISO, AF mode).

2. Go to **Menu → Setup Menu → Save user settings**.

3. Choose **U1 or U2**.

4. Press **OK** to save.

Now, simply turn the mode dial to U1 or U2 to recall your favorite setup instantly.

Example Use:

- **U1:** Outdoor portraits (A mode, f/2.8, Auto ISO, Face AF)

- **U2:** Sports (S mode, 1/1000 sec, AF-C, burst mode)

TIP: You can overwrite and update custom modes anytime.

It's like saving your favorite radio stations.

CUSTOMIZING YOUR NIKON Z50

ASSIGNING BUTTONS

1. MPress   Fn2   AE-L / AF-L
   MENU   (White balance)   Subject tracking

Fn1/ISO
Sensitivity

ISO
(ISO Sensitivity)   Fn2 (ISO — White balance)
(AE-L / AF-L)   AE-L / AF-L) Subject tracking

3. Confirm desired func

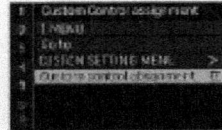

1. Go to CUSTOM MENU >
   CUSTOM CONTROL
   ASSIGNMENT
3. Select function

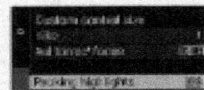

CREATING CUSTOM SHOOTING MODES

1. Choose mode (like A),
   adjust settings
2. Open Setup MENU >
   SAVE USER
   SETTINGS
3. Save to MODE DIAL
4. Choose U1 or U2

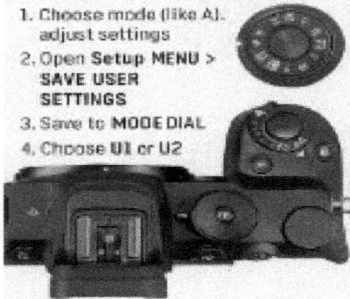

1. Set to BBN UA & SELE
2. Go to CUSTOMIZE i MENU
3. Turn FRAMING GRID ON

TURNING ON FOCUS PEAKING

1. Set to MANUAL
   FOCUS (MF)
2. Go to
   CUSTOM SETTINGS >
   PEAKING HIGHLIGHTS
3. Set funcing ON

TURNING ON GRID LINES

1. Press
2. Select CUSTOM
   MENU
3. Turn FRAMING
   GRID ON

Turning on Focus Peaking and Grid Lines for Clarity

When using **manual focus** or composing carefully, your

Z50 has built-in visual aids that help keep things sharp and straight.

Focus Peaking (Helps with Manual Focus)

Focus peaking outlines **in-focus edges** in bright colors (like red or yellow) so you can instantly tell what's sharp—especially helpful for close-ups or portraits.

How to Enable:

1. **Menu → Custom Settings → Peaking Highlights**
2. Set to **ON**
3. Choose a color (Red, Yellow, Blue, or White)

*Best used when in **Manual Focus (MF)** mode. You'll see outlines shimmer when an object is in perfect focus.*

Grid Lines (Helps with Composition)

Grid lines help you follow the **Rule of Thirds**, keep buildings straight, and align horizons. They're invisible in your final image but a huge help while composing.

How to Enable:

1. Press the **i Button** during shooting.

2. Choose **Display Options**.

3. Enable **Framing Grid**.

*TIP: Combine **grid lines with level indicators** for extra composition control.*

Summary:

- Assign **ISO, WB**, or **focus modes** to front buttons for quick changes.

- Save your favorite settings to **U1/U2** on the mode dial.

- Turn on **focus peaking** for crisp manual focus.

- Enable **grid lines** to straighten shots and improve your framing.

These tweaks make your camera **feel intuitive, faster**, and **personalized to your needs**—whether you're just learning or shooting like a pro.

Chapter 13

Storage, Cleaning, and Backup

Keep Your Camera Clean, Your Photos Safe, and Your Cards Ready

How to Safely Clean Your Camera Sensor and Lens

Proper cleaning protects your camera and image quality without needing expensive tools or a technician for every smudge.

1. Cleaning the Lens (Front Element)

Dust or fingerprints on the lens can make your photos appear hazy or blurry.

What you need:

- Microfiber cloth
- Lens cleaning solution or lens wipes
- Rocket blower

Step-by-Step:

1. Blow away loose dust using a **rocket blower** (never your breath).
2. Gently wipe the lens in a circular motion with a clean microfiber cloth.
3. If needed, apply a drop of lens cleaner to the cloth— never directly on the lens.

TIP: Always replace the lens cap when not using your camera to avoid dirt buildup.

2. Cleaning the Camera Sensor

Sensor cleaning is delicate. Only do it if you notice persistent dust spots in your images (especially visible in skies or light backgrounds).

What you need:

124

- Rocket blower (ONLY—never touch the sensor directly!)

Steps:

1. Remove the lens.

2. Hold the camera facing down.

3. Gently squeeze the blower to remove dust without touching the sensor.

4. Reattach the lens.

WARNING: If dust remains, visit a camera technician. **Never wipe the sensor** *unless you're trained.*

3. Wipe Down the Body

- Use a slightly damp microfiber cloth to clean buttons, screen, and grip areas.

- Avoid cleaning fluids near openings like ports or the lens mount.

Backing Up Your Photos to a

125

Computer or Cloud

Imagine losing an entire vacation's worth of photos because your SD card failed. Backing up is non-negotiable and easier than you think.

A. Backing Up to a Computer

Option 1: Using a Card Reader

1. Remove your SD card from the Z50.

2. Insert it into your computer's card reader.

3. Open the DCIM folder and copy all photos to your Pictures folder or external drive.

Option 2: Using a USB Cable

1. Connect your Nikon Z50 to your computer via USB.

2. Power on the camera.

3. Your camera will appear as a drive; drag and drop photos as needed.

TIP: Organize folders by date or event (e.g., "2025-05-TripToBanff").

B. Backing Up to Cloud Services

Popular Services:

- **Google Photos** (free up to a limit)
- **Dropbox**
- **iCloud**
- **Amazon Photos** (free for Prime users)

Step-by-Step (Google Photos Example):

1. Transfer photos to **your phone or computer**.
2. Install the **Google Photos app**.
3. Enable **Backup & Sync**.

Cloud backup protects your memories even if your camera, phone, or laptop is lost or damaged.

Formatting Memory Cards the Safe

Way

Formatting clears your memory card and prepares it for new photos. Do this after backing up, and before big shoots.

Why Format Instead of Just Deleting?

- Prevents file corruption
- Keeps the card running fast and error-free
- Deletes hidden system files that can slow down performance

How to Format Your SD Card:

1. Insert the memory card into your Z50.
2. Press the **MENU** button.
3. Navigate to **Setup Menu** (wrench icon).
4. Select **Format memory card.**
5. Choose **Yes** and press OK.

This will erase everything—so always back up first!

TIP: Avoid formatting your card on a computer—it's safer to do it in-camera.

Summary:

- Clean your lens gently with a blower and microfiber cloth. Never touch the sensor directly.

- Back up regularly to a computer and cloud to protect your memories.

- Format your SD card in the camera after every backup for smooth performance.

Keeping your Nikon Z50 clean and your photos backed up is as important as learning how to shoot. A clean, ready camera and a safe memory archive means you're always ready to create.

Chapter 14

Wi-Fi & SnapBridge Made Simple

Easily Transfer Photos, Control Your Camera Remotely, and Fix Connection Glitches

Wi-Fi & SnapBridge Made Simple

Transferring Photos to Your Smartphone
1. Install the SnapBridge app
2. MENU > Connect to smart device
3. Pair the camera with your smartphone

Download photos

Using SnapBridge for Remote Shooting
1. Open SnapBridge app
2. Tap "Remote photography"
3. Use your phone to focus and shoot

Control camera wirelessly

Common Connection Errors and How to Fix Them

Camera not found	Photos aren't transferring	Remote shooting won't work
Restart both devices and retry	Verify permissions and restart the app	Turn on Wi-Fi in camera menu

Transferring Photos to Your Smartphone

You don't need cables or a computer to share your Nikon Z50 images. With SnapBridge, you can send your photos wirelessly to your smartphone—ready for saving, printing, or sharing online.

What You Need:

- A **smartphone or tablet** (iOS or Android)
- The **SnapBridge app** (free from App Store or Google Play)
- Bluetooth and Wi-Fi enabled on both devices

Step-by-Step: First-Time Setup

1. **Install SnapBridge** on your phone.
2. Turn on your **Nikon Z50**.
3. Go to **Menu → Setup Menu → Connect to Smart Device.**

131

4. Choose **Start pairing**.

5. On your phone, **open SnapBridge** and follow the pairing instructions:

   - Allow all permissions (Bluetooth, Photos, Location).

   - Select your camera model when it appears.

6. Confirm the **matching PIN code** on both the camera and phone.

7. Once paired, choose if you want **Auto Image Transfer ON or OFF.**

TIP: You only need to pair once. After that, it connects automatically whenever both devices are on and nearby.

Transferring Photos

Once paired:

- Open SnapBridge.

- Tap **"Download Pictures"**.

- Choose the images you want to transfer.

- They appear in your phone's camera roll/gallery instantly.

Auto transfer will send photos to your phone every time you take a shot—great for backing up daily photos without manual effort.

Using SnapBridge for Remote Shooting

SnapBridge lets you control **your camera wirelessly** from your phone. This is perfect for:

- Self-portraits and group photos (no need for a timer!)
- Wildlife photography (stay at a distance)
- Tripod shooting (no camera shake)

How to Use Remote Shooting:

- Ensure your phone is still connected to the Z50.

133

- Open SnapBridge and tap "**Remote Photography**."

- Your camera's live view appears on your phone.

- Tap to **focus**, then press the shutter button **on your phone screen**.

- Your image is saved to the camera and optionally sent to your phone.

*Use this for **hands-free vlogging**, family portraits, and even **low-light shooting** where touching the shutter might cause blur.*

Common Connection Errors and How to Fix Them

If your SnapBridge connection is failing or seems buggy, don't worry it's usually fixable in a minute or two.

Problem: "Camera Not Found" or Doesn't Pair

Fix:

- Make sure **Bluetooth is turned ON** for both devices.

- Restart both the phone and the camera.

- Unpair from your phone's Bluetooth settings and **re-pair from scratch**.

Problem: Photos Aren't Transferring

Fix:

- Check if SnapBridge has **permission to access** photos on your phone.

- Ensure the camera's **Auto Image Transfer is ON**.

- Verify you're not in **Airplane Mode** or on a weak Wi-Fi signal.

Problem: Remote Photography Button is Grayed Out

Fix:

- Make sure your **Wi-Fi** is ON.

- Camera must be **not in video mode** or deep settings menu.

- Disable any **power-saving modes** on your phone that block background connections.

Summary:

- **SnapBridge** connects your Z50 to your phone via Bluetooth and Wi-Fi.

- You can **transfer photos**, **control the shutter remotely**, and **back up to your phone** effortlessly.

- Connection issues are usually simple to solve—restarting and re-pairing fixes 90% of problems.

Chapter 15

Troubleshooting Common Issues

*Simple Fixes for the Nikon Z50—No Tech Knowledge
Needed*

Camera Not Turning On? Here's
What to Check.

It's a scary moment, you press the power switch and...
nothing. Don't panic. It's usually something small.

Step-by-Step Checklist:

1. Battery Inserted Correctly?

- Open the battery door and check if the battery is
  seated firmly with the contacts facing the correct
  direction.

2. Battery Charged?

- Plug it into the charger and charge until the light turns off (fully charged).
- Try a second battery if available.

3. Power Switch Flipped Properly?

- Make sure the On/Off ring around the shutter button is turned fully clockwise.

4. Memory Card Inserted?

- Some cameras may not fully start up without a card in place.

5. Frozen System?

- Remove the battery for 10 seconds and reinsert it. This resets the system.

TIP: If none of this works, try resetting the camera settings from the Menu or contact Nikon support for diagnostics.

Autofocus Not Working? Try These Fixes.

If your camera won't focus, it might not be broken—it's likely just a setting or condition that's confusing the autofocus.

Common Fixes:

1. Lens Fully Attached?

- Remove and reattach the lens until it clicks.

2. Switch Focus Mode to AF

- On-screen: Tap the **"i" button** → **Focus Mode** → **Select AF-S or AF-C.**
- Or check for a physical switch on the lens and set it to **A** (not M).

3. Clean the Lens Contacts

- Gently wipe the gold contacts on the lens and inside the camera with a dry microfiber cloth.

4. Light Too Low or No Contrast?

- The Z50 needs light and edges to focus. Try aiming at something with texture or turning on more light.

5. Face/Eye Detection Disabled?

Press the "i" button → AF-area Mode → Face/Eye Detection ON

TIP: Use AF-S for still subjects, AF-C for moving subjects.

Why Are Your Photos Blurry or Too Dark? Solved.

Even if your Z50 is working perfectly, settings and technique can make photos look "off."

Blurry Photos:

Problem	Fix
Camera shake	Use faster shutter speed (1/125 or more), enable VR
Subject moved	Switch to AF-C and use 1/1000 shutter speed
Focus missed	Use Single Point AF and tap to focus before shooting
Low light	Raise ISO (800–1600) or use more light

Dark Photos:

Problem	Fix
ISO too low	Increase to ISO 400–800
Fast shutter speed	Use 1/60 or slower for more light

Scene backlit	Use Exposure Compensation (+1.0) or add light to the front
Lens aperture too narrow	Use wider aperture (f/3.5– f/5.6)

TIP: Use the Playback Zoom button to review focus after each shot. Zoom into the eyes of your subject to confirm sharpness.

Memory Card Problems: Formatting, Errors, Recovery

"Card Cannot Be Used" or "Card Error" Appears?

1. Try Re-seating the Card

- Remove and reinsert it firmly.

2. Format in Camera

- MENU → Setup Menu → Format memory card → OK

- **WARNING**: This deletes all images—only do this after backing up.

3. Try a Different Card

- Cards can go bad over time. Use **UHS-I SD** cards from brands like SanDisk, Lexar, or Samsung.

4. Card Locked?

- Slide the small switch on the side of the SD card **up** (toward the contacts).

Recovering Lost Images

If you deleted images by mistake or your card was formatted before backing up:

- **Stop using the card immediately.**
- Use recovery software like Recuva, EaseUS Data Recovery, or PhotoRec.

143

- Insert the card into a computer and follow recovery tool instructions.

Recovery success is higher if you haven't written new data to the card.

Final Troubleshooting Tips:

- **Reset your camera settings** to factory defaults via Menu if everything feels off.

- **Always carry a spare battery and memory card** for peace of mind.

- Don't ignore signs like flickering screens, freezing, or unexpected shutdowns—they usually signal a simple fix.

PART VI: Bonus Tips, Inspiration & Next Steps

Chapter 16

Hidden Features & Shortcuts You'll Love

Discover the Quiet, Clever, and Surprisingly Useful Tools Built into Your Nikon Z50

Silent Shooting Mode

Want to take photos quietly without the usual shutter sound? Whether you're at a wedding, museum, wildlife reserve, or trying not to wake a sleeping baby, **Silent Photography** is a life-saver.

What It Does:

- Eliminates the shutter sound entirely
- Uses an electronic shutter (no physical click)
- Ideal for quiet environments or discreet shooting

How to Enable:

1. Press the **i Button**.

2. Tap **Silent Photography** and switch it ON.

TIP: It works best in bright light. In very low light or with fast movement, images may blur so it's ideal for still subjects in quiet scenes.

Eye Detection Autofocus (AF)

This is one of the Z50's most powerful (and underrated) features. When enabled, your camera automatically **detects faces and locks focus on the eyes** ensuring pin-sharp portraits every time.

Perfect For:

- People photography
- Children or group shots
- Vlogging and video interviews

How to Use It:

1. Set Focus Mode to **AF-S** (still) or **AF-F** (video).

2. Go to **AF-area Mode** → **Auto-area AF** (People).

3. The camera will automatically identify faces and track their eyes.

You'll see a **yellow box** around the eye, press the shutter and enjoy razor-sharp results.

TIP: Works with both stills and video. Combine with Silent Mode for natural, unobtrusive portraits.

Self-Timer Tricks

The self-timer isn't just for group photos. Use it to eliminate camera shake, get in the shot yourself, or even create motion blur effects with no physical contact.

How to Set It:

1. Press the **i Button**.

2. Select **Release Mode** → **Self-Timer**.

3. Choose **2s, 5s, 10s, or multiple shot delay**.

Creative Uses:

- **2s Timer:** For tripod shots to reduce shake

- **10s Timer:** Join group photos

- **Multiple Shots:** Take 3–9 images in a row—perfect for capturing just the right moment

TIP: For best results, combine self-timer with VR ON and burst mode for sharp, steady photos.

Using the Customizable iMenu

The **i Menu** is your Z50's **fast-access control panel** and you can personalize it to match your shooting style.

Instead of hunting through menus, place your most-used settings just a tap away.

How to Customize:

1. Press the **MENU** button.

2. Go to **Custom Settings Menu (pencil icon)**.

3. Select **f1: Customize i Menu**.

4. Choose which functions appear in your i Menu grid:

    - Silent Shooting

    - Focus Mode

    - AF-Area Mode

    - White Balance

    - Picture Control

    - ISO Sensitivity

    - Image Quality

    - Grid Display

    - Exposure Compensation

    - Wi-Fi Control

Choose only the tools you use regularly, don't overload it!

Pro Shortcut Tip:

You can also **customize the Fn1 and Fn2** buttons to access your favorite iMenu settings without ever touching the screen.

Go to:

MENU → Custom Settings → Controls → Custom control assignment

Assign things like ISO, White Balance, or Eye Detection to the buttons by the lens for **one-handed operation**.

Summary:

- **Silent Mode** makes your Z50 stealthy and discreet.
- **Eye Detection AF** takes your portraits to a professional level.
- **Self-Timer** eliminates shake and captures perfect candid shots.
- **The i Menu** puts power at your fingertips— customize it to shoot faster, smarter, and more comfortably.

These small features have a big impact on how enjoyable, seamless, and professional your experience with the Nikon

Z50 becomes.

.

Chapter 17

Photography Tips for Seniors & Visual Learners

See More Clearly, Press More Comfortably, and Shoot Without Stress

Adjusting Brightness and Font Size on the Camera Screen

For many seniors or users with lower vision, the default settings can feel too dim or text too small. The good news? The Nikon Z50 offers ways to brighten your screen and improve readability.

A. Increase Screen Brightness:

1. Press the **MENU** button.

2. Go to **Setup Menu** (wrench icon).

3. Scroll to **Monitor Brightness**.

4. Adjust to **+1 or +2** for outdoor use or brighter visibility.

TIP: The Electronic Viewfinder (EVF) has a separate brightness control too—adjust it under Viewfinder Brightness in the same menu.

B. Improving Screen Readability:

While you can't enlarge menu fonts directly, you can improve visibility by:

- **Turning on grid lines** for clear visual spacing.
- Using **high-contrast menus** (white text on black background).
- Turning off excess info displays (simplifies the screen view).

Bonus Setting: Simplified Info Display

1. Press the **DISP** button while shooting.
2. Cycle through display modes until you find the **simplest layout**.

154

Use this mode to eliminate distractions and keep the focus on your photo, not the technical settings.

Using Larger Buttons and Grips

The Nikon Z50 is already known for its **comfortable grip and compact design**, but you can still make it **even easier to handle** with a few simple accessories and ergonomic tricks.

A. Add a Grip or Thumb Rest

- Use **an aftermarket silicone or metal hand grip extension** for more surface to hold.

- A **thumb rest attachment** improves stability and prevents slips.

B. Use a Soft Shutter Button

- Add a **soft-release shutter button** for a bigger, more responsive feel.

- Makes it easier to press without accidentally shaking the camera.

C. Use a Comfortable Strap System

Replace the neck strap with a wrist strap or crossbody sling strap for lighter carrying and easier access.

Pro Holding Tip for Seniors:

- Use your **right hand to grip** the camera and **left hand to support** the lens underneath.
- **Tuck your elbows** in to your sides, this stabilizes your arms and reduces blur from hand shake.

Setting Up Easy Mode for No-Fuss Shooting

The Z50 doesn't have a one-button "Easy Mode," but you can build one using these settings for an **effortless, frustration-free shooting setup.**

Step-by-Step: Create Your Own Easy Mode

1. Turn the **mode dial to AUTO** (green icon).

2. Press the **MENU** button and set:

- **AF-Area Mode** → Auto-Area

- **Focus Mode** → AF-S (still) or AF-F (video)

- **Auto ISO** → ON

- **Silent Photography** → ON (optional)

- **Beep** → ON (audio cue for focus lock)

3. In **Custom Settings** → **f1: Customize i Menu**:

- Add only the essentials: ISO, White Balance, Silent Mode, Picture Control

4. In **Custom Settings** → **Assign Fn1/Fn2 Buttons**:

- Fn1 = ISO

- Fn2 = Flash ON/OFF or White Balance

5. Save these settings as U1 or U2 (via Setup Menu → Save User Settings).

Next time, just turn the dial to U1—and you're instantly in your custom "Easy Mode."

Optional: Add a Large Screen Viewfinder Loupe

A magnifying loupe attaches to the screen, giving a bigger and clearer view for framing and focusing. Ideal for anyone with visual impairment.

Summary:

- **Brighten the screen** and use simplified displays to make menus more readable.

- **Add ergonomic grips, straps, and soft buttons** for better comfort.

- Create a **custom U1 Easy Mode** that handles settings automatically—just turn it on and shoot.

- Don't let tech slow you down—your Nikon Z50 is more adaptable than you think.

Chapter 18

Creative Photography Ideas to Try Today

Fun, Personal Projects to Build Confidence, Skill, and Memories

1. Photo-a-Day Challenge

A simple yet powerful way to grow your skills and start seeing the world differently.

How It Works:

Take **one photo every day**—no rules, no pressure. Just observe, compose, and capture something meaningful.

Ideas to Try:

- Your breakfast or tea
- What's outside your window
- A flower in bloom
- Shadows on the wall

- Something that made you smile

TIP: Choose a theme each week: textures, colors, light, movement, or emotions.

Why It's Great:

Builds a daily habit without overwhelm

Helps you practice framing and lighting

Creates a visual journal of your life

You don't need to post them online—keep a **private album or print them monthly**.

2. Nature Walks and Journaling

Pair photography with mindfulness by turning your morning walk into a creative expedition.

What to Bring:

- Your Nikon Z50 with kit lens
- A small notebook or voice recorder (optional)
- Comfortable shoes and an open mind

What to Photograph:

- Dew on leaves or spider webs

- Birds, trees, mushrooms, or footprints

- Cloud patterns or tree shadows

- The same trail each week, watching it change

*Combine photography with a short **written reflection**: what did you notice? How did it feel?*

Why It's Great:

- Combines movement, creativity, and mental health

- Helps you slow down and notice beauty in small places

- Builds a photo series that evolves with the seasons

3. Photographing Your Grandchildren

Whether at play or during quiet moments, these photos become treasured memories.

Poses and Ideas:

- Laughing candid shots during playtime

- Reading a book together

- Baking in the kitchen or playing with pets

- Silhouettes at sunset, running outdoors

Camera Tips:

- Use **Shutter Priority (S Mode)** at 1/250 or faster to freeze motion.

- Switch to **Continuous AF (AF-C)** for moving kids.

- Use **Portrait Scene Mode** for soft backgrounds.

For natural expressions, don't say "Smile." Say something funny or unexpected.

Bonus: Let them take a photo of you too—it teaches them and creates shared moments.

4. Holiday and Travel Storytelling Through Pictures

Even if you're just taking a day trip or hosting family at home, there's a story to be told.

Tell the Story in 5 Shots:

1. **The Setting** — wide shot of where you are

2. **The Details** — close-ups of food, signs, hands, or decor

3. **The Faces** — capture joy, focus, and interaction

4. **The Action** — kids playing, conversations, walking, etc.

5. **The Memory** — a quiet moment, sunset, or goodbye

Creative Formats:

- Make a **travel photo book** (services like Chatbooks or Shutterfly)

- Create a **slideshow with music** for family

- Print a "**10 moments from our trip**" **collage**

Bonus idea: ***use a custom folder*** *or* ***tag in your camera*** *to separate special photo projects by theme or event.*

Photography Ideas

Photo-a-Day Challenge
- Take one photo every day
- Capture small moments in life
- Explore themes like colors shadows

Nature Walks & Journaling
- Take a camera on walks
- Photograph flowers, birds or trees
- Jott down thoughts in a journal

Photographing Grandchildren
- Document playtime, everyday activities
- Focus on natural smiles & interactions
- Use a fast shutter speed for active kids

Holiday & Travel Storytelling Through Photos
- Capture your sights and loved ones on trips
- Show details, scenes, & special moments
- Create a visual story of your journey

Final Thought:

You don't need exotic places, professional skills, or fancy equipment. Just one camera, your Nikon Z50, and a little **intention**.

Make your photography personal. Let your photos say,

"I was here. I saw this. It mattered."

Glossary of Essential Photography Terms (Plain English)

Term	Meaning
Aperture (f-stop)	How wide the lens opens—affects brightness and background blur
Shutter Speed	How fast the photo is taken—controls motion blur
ISO	Camera's sensitivity to light—higher = brighter (and more grain)
Exposure	The brightness of a photo—controlled by aperture, shutter, ISO
Depth of Field	How much of the image is in focus (shallow = blurry background)

White Balance	Adjusts color based on lighting (sun, shade, bulbs)
Autofocus (AF)	Camera automatically focuses on a subject
Manual Focus (MF)	You focus by turning the lens ring
Burst Mode	Takes multiple photos rapidly when you hold the shutter
EVF (Electronic Viewfinder)	Small screen you look through instead of the rear LCD
Live View	Shooting with the LCD screen instead of the EVF
Scene Modes	Pre-set camera settings for specific situations like portraits or landscapes

RAW vs JPEG	RAW = large, editable file; JPEG = smaller, ready-to-use file
VR (Vibration Reduction)	Helps reduce blur from hand movement
SnapBridge	Nikon's app to transfer photos and control the camera via smartphone
Peaking Highlights	Shows what's in focus when using manual focus

Acknowledgement

To every aspiring photographer and filmmaker who dares to pick up a camera and tell a story, this book is for you. Special thanks to my family and friends for their encouragement, and to the creative community whose passion inspires me daily. Your support made this guide possible.